I0423000

Requiem for Humanity

ALAN HOSHOR

Published 2011 by Alan Hoshor

Requiem For Humanity. Copyright © 2011 by Alan B. Hoshor.

All rights reserved. No part of this publication may be reproduced, stored in a retrieval system, or transmitted in any form or by any means, digital, electronic, mechanical, photocopying, recording, or otherwise, or conveyed via the Internet or a Web site without prior written permission of the publisher, except in the case of a brief quotation embodied in critical articles and reviews.

Inquiries should be addressed to:
Alan Hoshor
info@requiemforhumanity.com
www.requiemforhumanity.com

ISBN: 978-1463568726
LCCN: 2011905839

ACKNOWLEDGEMENTS

This book would never have been written without the encouragement of three people: my wife, Ginger Hoshor; my father, James Hoshor; and my friend, Bill Mitchell (AKA Boracay Bill). My prose skills are dismal and writing is a tremendous struggle for me. These three recognized my strength in systems thinking and pushed me to share my insights with a broader audience. Another lifelong friend, Mary Klement, worked with me through the entire manuscript discussing each concept and validating that my prose was communicating what I intended. Cover design is the work of Jennifer Eichmeier who helped me create a visual metaphor for my book.

CONTENTS

INTRODUCTION

The future of humanity is seen by many to be on the cusp of change. Those who predict the future see tomorrow through a lens formed from their education, experience and viewpoints. Bill Joy presages an apocalypse; Ray Kurzweil expects to ride a technologic stairway to heaven; and Jaron Lanier predicts humanity will transcend itself into an unforeseeable future. This book sees the future through the lens of a generalist. I apply critical thinking and realistic application of diverse science into my prophecy for humanity over this century. My hope is to give the general reader a rational context for viewing the changes transforming our world today. It is better to know where the tiger we are riding is headed—if we hope to hold on and maybe even steer.

1 PROLOG

Humanity is in the midst of the single most dramatic change that Homo sapiens (our species) will ever experience. The human race is heading toward trouble. The trouble is of our own making and is unavoidable. Our primary crisis is that we have deduced the biology of genetics. Humanity is in the first steps of self-modifying our own species. This is unique on Earth; a living organism that can use its intelligence to redesign itself. No longer are thousands of years of evolution required. We will be modifying our own genome in one human generation.

This book is a collection of essays that attempt a balanced view of various topics that will influence our next century. There are many futurologists from the twentieth century that have shown deep insight into our future. I quote references from many of them in this prolog and throughout this book. I have included no bibliography and do not present my work as one of scientific research. Most of the ideas I present are best articulated by their referenced authors. What is unique to this book is that it is intended to persuade a variety of readers that our children are going to have a more dramatic influence on the destiny of humanity than any past or future generation. Much of what I have written here will cause your common sense to rebel. I ask only that you judge for yourself the path that humanity is on so that the choices we collectively make in the twenty-first century are as informed as possible. Most futurists are in agreement that the human experience will change more in the next century than in the previous 10,000 years. I try to realistically evaluate these forces for change and project some of the highly probable consequences to humanity.

For those readers familiar with historical predictions of impending calamity; they will recognize that such warnings have been a part of all recorded history. In the middle ages, the cross-bow threatened the end; later came the invention of dynamite; most recently, nuclear weapons threatened cataclysm. Still humankind has managed to survive and prosper. Many futurists believe we are currently facing what Vernor Vinge is credited as defining as a technological singularity. The consequences of this singularity on humanity are hotly debated. In his thoughtful book, *Radical Evolution – The Promise and Peril of Enhancing Our Minds, Our Bodies—and What It Means to Be Human*, Joel Garreau has termed the prospect of humanity muddling through 'The Prevail Scenario'. Joel extensively quotes Jaron Lanier's skepticism about the idea of a Singularity—technology increasing so quickly as to create an imminent and cataclysmic upheaval in human affairs. As you read farther in this book you will have to form your own opinions about humanity's robustness. I offer my thoughts, not as warnings, but to foster self-awareness; at best, to provide insight into the turbulent times ahead of us.

Jared Diamond, Pulitzer Prize-winner for *Guns, Germs and Steel* has written a contemporary book titled *Collapse: How Societies Choose to Fail or Succeed.* Jared is a multi-disciplinary biologist and historian. He peers beneath traditional cultural-historical explanations for the failure of past societies and searches for underlying fundamental causes. Diamond wrote that when our human ancestors choose to abandon the hunter-gather life style by inventing agriculture, the result was both rapid overpopulation and depletion of resources. In *Collapse*, Diamond warns that the lifestyle of current human societies has less than 50 years left.

I agree with Diamond that in this century humanity will experience a dramatic increase in calamities compared to our comparatively stable 20th century. Diamond considers the invention of agriculture the worst mistake in human history. I see it instead as a turning point that has led us to the cusp of control over our own evolution. *Collapse* has a fleeting few pages addressing new technology. Jared Diamond views technology as neither a primary threat nor source of solution for the modern world. In contrast I see the success of genetics and artificial intelligence, as game-changers which will vastly exceed the dramatic impact agriculture had on our hunter-gatherer ancestors. Modern futurist thinkers frequently view science and technology as either the path to human nirvana, or to the extinction of the human race. My view is akin to Jaron Lanier's; that we are beginning a frightful, rapid and messy evolution into multiple new intelligent species. These new species will incorporate machine intelligence, nanotechnology and organic biology.

Francis Fukuyama writes on the last page of his book *Our Posthuman Future - Consequences of the Biotechnology Revolution*:

> *"Many assume that the posthuman world will look pretty much like our own—free, equal, longer lives, and perhaps more intelligent than today. But the posthuman world could be one that is far more hierarchical and competitive than the one that currently exists and full of social conflict as a result. It could be one in which any notion of 'shared humanity' is lost, because we have mixed human genes with those of so many other species that we no longer have a clear idea of what a human being is. It could be one in which the median person is living well into his or her second century, sitting in a nursing home hoping for an unattainable death. Or it could be the kind of soft tyranny envisioned in Brave New World, in which everyone is healthy and happy but has forgotten the meaning of hope, fear, or struggle."*

Fukuyama's hope was to entreat humanity to avoid these potential outcomes. He asks our political communities to work together to protect the values which we consider to be innate human rights. This book takes the view that some isolated communities will successfully restrict the impact of artificial intelligence, genetic engineering, nanotechnology and robotics. Unfortunately they will not deter other communities from continuous rapid progress. The most successful of these communities will rapidly dominate our world.

Biological evolution has always been fraught with drama. We are in the process of inventing earth based alien life forms. Today people debate how close computers are to human intelligence. Replacing this debate, in the next century people will instead be alarmed by the new Technology Augmented Bioengineered (TAB) humans in endless competition for domination. In the following centuries no one will consider engineered intelligent life to be human. The human race is rapidly inventing its replacement. I believe that in five hundred years the only human beings on planet Earth will be in large preserves; isolated and protected like we do today with animals extinct in the wild.

2 PREDESTINATION

Is it human free will that will bring threatening technologies to life? Our social systems are designed to maintain the status quo. Those in power fight to maintain their authority. Societies that maintain cohesion under external threat are more likely to withstand disasters. Our leaders guide us against threats by invoking powerful human emotions like faith, self-protection, pride, and loyalty. Leaders don't try to invoke our higher logical reasoning powers. As a species we fall back on our instincts when stressed. This is a survival technique that has protected us for many thousands of years. We are not about to change. On this question I think the term 'human nature' is pertinent. Biologists currently believe that our genes control much of what we are. Our complex genetic makeup has imbued us with the capacity to risk our lives for our religion or nation. In the tumultuous time facing us, many groups will rationalize the need to develop artificial intelligence, genetic engineering, nanotechnology and robotics technologies to aid in medicine, and for economic/military survival. Today we see the polarized responses to issues like abortion or religious beliefs. Soon these will be submerged under the new issues of Technology Augmented

Bioengineered (TAB) humans. In 25 years, computer intelligence will exceed the smartest humans. In this century many people will have minds directly augmented by brain extensions wirelessly connected to centralized computers, their bodies amplified by drugs and genetic enhancements. This will further fracture humanity aggravating existing social strife.

George Friedman, the founder of Stratfor (world's leading private intelligence and forecasting company) has just published a book, *The Next 100 Years: A Forecast for the 21st Century*. On the topic of predestination, George Friedman writes that *"Free will is beyond forecasting. But what is most interesting about humans is how unfree they are." [...] "We are deeply constrained in what we do by the time and place in which we live."*

Biologist Anthony Cashmore, *The Lucretian swerve: The biological basis of human behavior and the criminal justice system*, argues that a belief in free will is akin to religious beliefs, since neither complies with the laws of the physical world. One of the basic premises of biology and biochemistry is that biological systems are nothing more than a bag of chemicals that obey chemical and physical laws.

So, here we are—bags of chemicals, only now these bags of chemicals are going to re-invent themselves. How do we escape our own genetic limits? The simple answer is that we won't. Existing social mores, religions, and political groups create fundamental limits to what kinds of near term change is possible. Inherent in our genetics are all the mechanisms evolution provided us for survival. Human individuals and tribal groups will fight to avoid change. Dramatic change will require social disruption and competition for diminished resources as incentives. As Jared Diamond has pointed out, the human species has a propensity to ignore the future consequences of present actions. So it will be the same as we begin bioengineering humans.

What are our next few decades going to be like? Our tomorrow is going to be very much like today. Social structures have too much inertia to change overnight.

Despite this inertia, society's hubris causes us to ignore early warning signs. Humans have a tendency to assimilate incremental change with little social disruption as we collectively ignore the future consequences of present actions. Our world will change around us. Most will welcome it. No better example of this can be found than the last 100 years. Luddite groups of the early 20th century did not slow the huge increases in population and the dramatic consumption of resources that challenge today's world. Similarly, the unbelievable compounding pace of change we live with today will far outpace efforts to contain or restrain it. Jared Diamond demonstrated that throughout history societies facing loss of critical resources were more likely to invoke deities or demonize neighbors than to deal rationally with their impending disaster.

Consider the Easter Island woodcutter portrayed in Jared Diamond's book *COLLAPSE*. In the chapter 'Twilight at Easter' an Easter Island native worker sweats to remove the last trees remaining on their island. The tree trunks needed to roll the final Moai stone statue to the ocean. It is inconceivable that same woodcutter was unaware that he was eliminating the last material available for constructing the fishing canoes which Easter's society depended on for survival. The island society chose faith in their priests and Polynesian gods over common sense. Modern societies are treating our current resource crisis in a similar fashion. It is human nature.

Bill Joy recently suggested (TED, Feb 2006) that society's best chance at survival is to attempt to guide our future, limiting the most dangerous potential paths to lower the probability of catastrophic risk. Joy is a genius computer scientist. He co-founded Sun Microsystems in 1982 and is widely known for having written the essay, *Why the future doesn't need us*, which suggested that development of modern technologies endanger the existence of life as we know it. While I agree with Bill about the risks to humanity, what he suggests is defensive in nature. History has shown that defenses always fall to prolonged attack and consume resources that lower the defender's economic health. Over generations, the primary survival technique is to 'out smart' your competitors. It is the nature of living things. It is our human nature. Governance will not be able to constrain the multitude of risks facing our modern world. It won't be individual free will that threatens the status quo of future societies. Competitions between countries for resources insure that they will develop risky technologies to gain advantage over their neighbors. This evolutionary escalation is the pattern of human history and is built into the behavior of living things.

In his landmark book, *Why the West Rules—For Now, The Patterns of History and What They Reveal about the Future*, Ian Morris submits that *"sloth, fear and greed are the motors of history."* He concludes that the world is at the precipice of disaster and that it can only be avoided by a unified world self-aware of its cyclic history of self-induced catastrophe. Like Bill Joy, Ian Morris doubts our chances for survival. I submit that it is the anarchic dynamism of individuals, not global governance, that pose our primary opportunity for humanity to evolve. The Morris Theorem, *"that change is caused by lazy, greedy, frightened people looking for easier, more profitable, and safer ways to do things"* is just another way of describing the evolutionary process. The principal process hasn't changed. Genetic engineering, nanotechnology and robotics are contributing more variables into mix. The pace of change is tremendously faster, but the competitive process of evolution stays the same.

3 HUMAN INSTINCTS

"The difference between men and women is that a woman wants one man to satisfy her every need, and a man wants every woman to satisfy his one need." Playboy's Party Jokes, November 2009. Most will see the humor intended by this joke, even if they don't like it being quoted from a chauvinist magazine like Playboy. The human female evolved to nurture the most genetically robust offspring possible. Contrasted with this, the human male is programmed to spread his genes as broadly as he can get away with. Early social mores developed to protect the female in her nurturing role and sent the male out to hunt and protect. It is little wonder with these dramatically different gender traits that modern societies are still grappling with sexual equality. In some communities today's woman seems to have the best deal, having been accepted in both her biological role and offered social equality. In contrast, the males have diminished roles and their genetic programming to spread their genes is frowned on by church and state. Some feminist organizations even question the need for males in the future of the human race.

This leads to me to the question: "What are humans without our instincts?" Our instinctual responses to our environment are hard coded into our genome to protect us

9

while we lived in the natural world of our ancestors. So much of current society's structure is still in response to these instincts. Sports, cosmetics, fashion, laws, computer games, racial tension and fast cars all are based on fundamental instinctual drives. Sex is by far the most visible instinct in the modern world. It pervades everything. Societal responses to the human sexual drive are fundamental to religions, laws, and economies. It continuously impacts our lives on multiple levels.

With control of our own genome, future architects of 'purpose built' humans won't want the complication of extraneous instincts. If designing a warrior you might like loyalty, fearlessness, and aggression. You would definitely not want compassion, independence, or sexual drive. Lower brain emotions evolved from our animal ancestors. Modern cognitive science and genetics have recently discovered that morality is hard wired into our brain. The fundamental basis of all human behavior is to reproduce and to protect our offspring until they reach reproductive age. The human sexual drive dominates every aspect of modern civilization and it soon will have no purpose! What will it be like? What will the motivations be of future humans? Without our sex drive what happens to the institution of marriage? If woman no longer need men as fathers and neither men nor women are motivated by sex, why would we ever marry? Wirelessly connected in a hive conscious; will our societies be more like those of ants or bees? Popular press wonders what robots will think. I wonder what we modified humans will think.

4 ANARCHY AND EVOLUTION

Competition of life forms (bio-human, android, nanobots, robot) is critical for them to evolve into robust species capable of dealing with all the diversity in the universe. Early earth life had millions of years for this competition. Science has demonstrated that co-evolutionary interactions with other species are the primary cause for rapid evolution. During the singularity, we humans will be diverging into competing life forms.

From subatomic particles to intergalactic space, the universe is in constant competition for resources. A tree shades its neighbor, stealing sunlight; ants and aphids conspire to suck the tree's sap from its phloem vessels; the aphids' honeydew takes the solar energy converted by photosynthesis providing carbohydrates to the ant-hill. The ant-hill's extensive tunneling system damaging the home garden and consuming resources that would have provided calories from the sun to you and me. This complex interdependency of Earth life evolved over millions of years.

Research in 2010 published in the journal *Nature* credits scientists at the University of Liverpool as providing the first experimental evidence that shows evolution is driven most powerfully by interactions between species, rather than adaptation to the environment. The team observed viruses as they evolved over hundreds of generations to infect bacteria. They found that when the bacteria could evolve defenses, the viruses evolved at a quicker rate and generated greater diversity, compared to situations where the bacteria were unable to adapt to the viral infection.

The theory, first put forward in the 1970s, was named after a passage in Lewis Carroll's *Through the Looking Glass* in which the Red Queen tells Alice, *"It takes all the running you can do to keep in the same place."* This suggested that species are in a constant race for survival and must continue to evolve new ways of defending themselves throughout time.

Dr. Steve Paterson, from the University's School of Biosciences, explains: *"Historically, it was assumed that most evolution was driven by a need to adapt to the environment or habitat."* The Red Queen Hypothesis challenged this by pointing out that actually most natural selection will arise from co-evolutionary interactions with other species, not from interactions with the environment. What this implies for our future melding of synthetic organisms and living machines is that to be robust species, they must have evolved with unconstrained competition for shared resources. It is the nightmare scenario that haunts Bill Joy. It is also very likely. Once intelligent machines are designing themselves, their goals will no longer be human based. They likely will simulate new machine designs with internal models, and then generate machine prototypes instrumented to measure success in a real world co-evolutionary environment. The intelligent machine will then use this feedback in a constant cycle of machine design, simulation, prototype, and then machine/model update. It is very conceivable that this process could accomplish in weeks what natural biology based co-evolution required hundreds of thousands of years to accomplish.

Bill Joy's April 2000 *Wired* article quotes Eric Drexler's Engines of Creation:

"Plants" with "leaves" no more efficient than today's solar cells could out-compete real plants, crowding the biosphere with inedible foliage. Tough omnivorous "bacteria" could out-compete real bacteria: They could spread like blowing pollen, replicate swiftly, and reduce the biosphere to dust in a matter of days. Dangerous replicators could easily be too tough, small, and rapidly spreading to stop—at least if we make no preparation. We have trouble enough controlling viruses and fruit flies.

Among the cognoscenti of nanotechnology, this threat has become known as the "gray goo problem." Though masses of uncontrolled replicators need not be gray or gooey, the term "gray goo" emphasizes that replicators able to obliterate life might be less inspiring than a single species of crabgrass. They might be superior in an evolutionary sense, but this need not make them valuable.

Bill Joy says that the Gray Goo threat makes one thing perfectly clear; we cannot afford certain kinds of accidents with replicating assemblers. In his Feburary 2006 TED presentation, Bill ends his talk entreating watchers to promote national policies that *"Limit access to great and unbridled power"*. He speaks of the million-to-one leverage of an individual (or small group) misusing powerful technology against our society's cost to protect against them. My view is that the only way to avoid the danger of a monoculture 'Gray Goo' life form dominating the earth is to assure that the same principles that allow bacteria to rapidly develop defenses when faced with an evolving viral threat work also for bio-machine competition. In simple terms we need Grey Goo eating machines co-evolving with the goo 'bacteria'.

Is this likely to happen? Consider how future nation states will be in fierce competition to control the life forms that will replace Homo sapiens. I predict bio-wars late in this century. The resource in contention will no longer be petroleum, water or land; instead it will be control of the synthesized life components from which superior humans will evolve. Threat of a neighboring nation developing a superhuman generation twice as smart, augmented by self-healing nanotechnology, and linked by artificial intelligence computers; will foster society-wide fear. Nation states will fight to both steal and destroy such capability. This is nothing more than our genetic predisposition to compete. By this point in our future we will likely have already experienced many disasters caused by widespread famine, drought, limited nuclear and biological war. The disparity between members of the human race will then be greater than when Cro-Magnon coexisted with Neanderthal for 15,000 years in France. Within one hundred years the competition between modern man and 'future' man will be in earnest.

I argue that by 2100, the human race will have begun to stratify into different breeds, eventually into different species (if the term species even applies). We will have begun to merge synthetic organisms and living machines into purpose made life forms tailored for particular environments and functions; living under water, living in space habitats, living on the moon and Mars. These new life forms could be compared to the evolution of mammals from reptiles. The transition has already begun. The Astrobiology Magazine 2010 article —*Cyborgs Needed for Escape from Earth* quotes historian Roger Launius, who calls himself a cyborg for using medical equipment to enhance his own life, says *"the difficult question is knowing where to draw the line in transforming human biological systems to adapt to space."* While financial and ethical concerns may have held back cyborg research, Launius believes that society may have to engage in the cyborg debate again when space programs get closer to launching long-term deep space exploration missions. George Friedman argues that when we try to predict the future, common sense almost always betrays us. In considering the rapid evolution of our species, all of our instincts betray us. It is extremely difficult to imagine the consequences of the genetic engineering, nanotechnology and robotics processes now shaping our future.

In counter-point, there is Ray Kurzweil's single-minded plan to live forever in a magical distant future. His dream blinds him to the messy future he is actually going to live in. What possible utility would a future world have for digitally stored brains? Even if we only stored one percent; consider how much redundancy and for what purpose? It seems so silly to me. Like a technologist's alternative to heaven. I don't think anyone alive can predict what we will be evolving ourselves into. The question for us today is whether we wish to collectively guide this evolution, and if so, whether that is even possible.

5 DISEASE

In an October 17, 2005 New York Times Op-Ed article titled *Recipe for Destruction*, Ray Kurzweil and Bill Joy collaborated on the following article:

AFTER a decade of painstaking research, federal and university scientists have reconstructed the 1918 influenza virus that killed 50 million people worldwide. Like the flu viruses now raising alarm bells in Asia, the 1918 virus was a bird flu that jumped directly to humans, the scientists reported. To shed light on how the virus evolved, the United States Department of Health and Human Services published the full genome of the 1918 influenza virus on the Internet in the GenBank database.

This is extremely foolish. The genome is essentially the design of a weapon of mass destruction. No responsible scientist would advocate publishing precise designs for an atomic bomb, and in two ways revealing the sequence for the flu virus is even more dangerous.

First, it would be easier to create and release this highly destructive virus from the genetic data than it would be to build and detonate an atomic bomb given only its design, as you don't need rare raw materials like plutonium or enriched uranium. Synthesizing the virus from scratch would be difficult, but far from impossible. An easier approach would be to modify a conventional flu virus with the eight unique and now published genes of the 1918 killer virus.

Second, release of the virus would be far worse than an atomic bomb. Analyses have shown that the detonation of an atomic bomb in an American city could kill as many as one million people. Release of a highly communicable and deadly biological virus could kill tens of millions, with some estimates in the hundreds of millions.

A Science staff writer, Jocelyn Kaiser, said, "Both the authors and Science's editors acknowledge concerns that terrorists could, in theory, use the information to reconstruct the 1918 flu virus." And yet the journal required that the full genome sequence be made available on the GenBank database as a condition for publishing the paper.

Proponents of publishing this data point out that valuable insights have been gained from the virus's recreation. These insights could help scientists across the world detect and defend against future pandemics, including avian flu.

There are other approaches, however, to sharing the scientifically useful information. Specific insights - for example, that a key mutation noted in one gene may in part explain the virus's unusual virulence - could be published without disclosing the complete genetic recipe. The precise genome could potentially be shared with scientists with suitable security assurances.

We urgently need international agreements by scientific organizations to limit such publications and an international dialogue on the best approach to preventing recipes for weapons of mass destruction from falling into the wrong hands. Part of that discussion should concern the appropriate role of governments, scientists and their scientific societies, and industry.

We also need a new Manhattan Project to develop specific defenses against new biological viral threats, natural or human made. There are promising new technologies, like RNA interference, that could be harnessed. We need to put more stones on the defensive side of the scale.

We realize that calling for this genome to be "un-published" is a bit like trying to gather the horses back into the barn. Perhaps we *will be lucky this time, and we will indeed succeed in developing defenses for these killer flu viruses before they are needed. We should, however, treat the genetic sequences of pathological biological viruses with no less care than designs for nuclear weapons.*

This warning is the complete editorial from Ray Kurzweil and Bill Joy. Considering how differently they view technology's impact on our future it is remarkable that they are in such agreement on the dangerous potential for humanity's destruction by robust lab designed pathogens. Ian Morris terms this a *"Nightfall"* scenario.

"For the Singularity to win, we need to keep the dogs of war on a leash, manage global weirding, and see through a revolution in energy capture. Everything has to go right. For Nightfall to win only one thing needs go wrong. The odds look bad."

As I said in the preface, history is filled with doom scenarios. *Nightfall* is also a science fiction short story by Isaac Asimov published in 1968. The name refers to the coming of darkness to the people of a planet ordinarily illuminated at all times on all sides. Civil disorder breaks out; cities are destroyed in massive fires and civilization collapses, with the ashes of the fallen civilization and the competing groups trying to seize control. This famous story serves as an apt metaphor for other genetic engineering, nanotechnology and robotics 'Nightfall' scenarios we will discuss in the following chapters.

6 BIOLOGICAL WEAPONS

How will human nature steer our nations over the coming decades? What are we evolved to do? The fundamental purpose of a living thing is to gather resources and reproduce. Some animals develop societies where the fundamental survival instinct of the individual is subsumed into the goal of protecting the social group. Human development of cities and defensive warfare addresses group survival under external threat. Our offensive warfare fundamentally addresses accumulation of resources required for survival. If you peel away all the layers of social rationalizations, at the base you will find this essential: human societies will react to future social threats attempting to insure survival of their nation states.

In his book *The Rise and Fall of the Great Powers* Paul Kennedy predicts that *"no other state is likely to join the pentarchy of the United States, the USSR, China, Japan, and the EEC in the near future."* When he wrote this in 1987, the triple threat of robotics, genetic engineering, and nanotech didn't exist. Now, small nation states of limited wealth and resources can afford development of bioengineered genetic weapons. Statesmen worry about nuclear weapons in the hands of authoritarian regimes in Iran and North Korea. The advent of genetic engineering raises another destabilizing specter.

Nuclear weapons are very messy to use. They physically destroy their target's resources and poison the surrounding area with radioactivity. This does not mesh well with the primary goal of offensive warfare, which is to accumulate resources. Genetically modified biological warfare agents, on the other hand, fit this goal perfectly. Done optimally, biological weapons destroy your enemy and leave his resources untouched. Bill Joy worries that artificial intelligence and nanobots are similar threats. While I don't disagree, the barriers to entry are high for artificial intelligence and nanotech. Genetic engineered weaponized biological agents are here today. Laboratories working on biological weapons are easy to hide under the guise of medical research. They are even within the means of a sophisticated terrorist. The main practical issue is dispersal before the target mobilizes defenses. New developments in genetic engineering are already enabling new concepts in biologic weapon delivery. There will be a multitude of ways to stealthily infect your enemy target in the near future. Paul Kennedy's pentarchy of International power will be replaced by *"anarchic and competitive ... rivalries between nations"* ending the relative peaceful and prosperous history of the last century.

Not to single out small nations, another scenario would be a current religion wishing to selectively eliminate all unbelievers; thus keeping the world's resources available for the chosen. It could be planned as a two stage attack. First the religion would fund a successful vaccine to a modern worldwide malady like the common cold or AIDS. Secretly they would attach a non-functional component that would cause rapid death if turned on by a unique amino acid trigger. To demonstrate the religion's humanitarianism they would offer a free version of the vaccine and invent a religious reason why all believers had to use a more expensive (acceptable to their God) version. After a few years everyone in the world would be vaccinated. Then the second stage; the 'missing key' trigger amino acid which causes death would be attached to a mild but hyper-virulent version of the flu. This flu would be simultaneously released worldwide. By the time that the world recognized that practitioners of the religion were immune most of the unbelievers would be dead.

Another scenario is described in the Tom Clancy novel, *Rainbow Six*, which involves a plot to unleash an engineered virus on humanity by infecting Olympic Games attendees with a virus carefully engineered to maximize virulence and lethality to the human host after enough time to infect others, in order to wipe out humanity except for the chosen few who had been immunized.

Large nations have the most resources available and would not likely be able to wait years or keep such biological weapons secret. Here a more likely scenario is dispersal of weaponized plague or flesh eating bacteria by a multitude of dispersal agents including animal vectors, autonomous aerial release, and water supply injection.

If these scenarios seem farfetched; consider that the Holocaust was only 55 years ago. A number of current societies are similarly xenophobic. There are both Muslim and Western groups that deny the Holocaust occurred, calling it a Jewish conspiracy. If Hitler could have infected kosher food with a delayed release non-infectious virus he would have used it. If historical revisionists can successfully guide their adherents to believe a re-write of one of the most despicable periods in modern history... what else can they rationalize? After World War Two, Eric Hoffer wrote in his preface to *THE TRUE BELIEVER Thoughts on the Nature of Mass Movements*,

> *"All mass movements generate in their adherents a readiness to die and a proclivity for united action; all of them, irrespective of the doctrine they preach and the program they project, breed fanaticism, enthusiasm, fervent hope, hatred and intolerance; all of them are*

capable of releasing a powerful flow of activity in certain departments of life; all of them demand blind faith and single hearted allegiance."

Mass movements arise not among the impoverished, but instead among relatively well-off middle classes facing loss of what they view as their entitlements. It is this middle class group that most easily yields their allegiance to a fanatic.

7 ECOLOGY

Humans are the most successful consumer of resources that Earth has produced. Our success has altered our planet. Humanity is currently experiencing limits to further growth. We have consumed the easily developed resources across most of our world. We are now faced with more people consuming fewer resources at a higher relative cost. Unchecked this trend will create the same results as in past human history: famine, war and disease.

The combination of human population explosion, global warming, commercial fish depletion, ocean acidification, depletion of underground aquifers, top soil erosion, and global resource consumption insure that human civilization will be collectively stressed. By 2050 we will have 50% more people, most of them in teaming cities; the commercial ocean fisheries will have collapsed eliminating the fish protein more than one billion depend on; huge expanses of the bread basket crop lands will no longer have water available to grow grains; fresh water will be limited in the Mediterranean, southwestern North America, north and south Africa, the Middle East, central Asia and India, northern China, Australia, Chile, and eastern Brazil. Less food, less water, and more people crowded into ever larger cities insure that global quality of life will be in dramatic decline.

In the 1970s, I was taught the first and second laws of thermodynamics. These theorems work really well in the ideal gas law and information theory. Thermodynamics explain how systems large and small tend to increase the total entropy of the universe. I generalized the concept in my own mental simplification that assumed all things in the universe are decomposing to their simplest elements. Quantum thermodynamics was beyond my grade level so I did not yet know that classic thermodynamics does not apply at the subatomic level. As I revisit my old simplistic assumption; I observe that the stars making up our universe start as a collection of simple gasses and then mature to create all of our heavy elements. An evolutionary model of our universe would show the stars which comprise it evolving from the simple to the complex and eventually collapsing back to the simple. Likewise, life's natural evolution generally favors the complex organism over its simpler predecessor in the endless battle for resources. Limited resources seems a fundamental universal constraint from the sub-atomic to the galactic.

India has 17% of the world's population and yet it currently has access to only 4% of the world's fresh water reserves. India's modern industry and population growth consume water at unsustainable rates. A US report, based on findings from NASA satellites, show that from 2002 to 2008, the volume of ground water available in Indian aquifers has decreased by 26 cubic miles, apparently due to the aquifers being drained faster than nature is replenishing them. Despite two giant groundwater reservoirs recently discovered in the Qaidam Basin, China's North China Plain region is rapidly losing its water reserves. Similarly, many locations in the United States are already undergoing water stress. Water stress created by the combination of ecological damage, population growth, industrial growth, and climate change is guaranteed to mean large communities are soon to face inadequate water supplies in the near future. The nonprofit Population Reference Bureau (prb.org) summarizes the status of the Middle East and North Africa:

"The challenge of addressing freshwater shortages in MENA [Middle East and North Africa] is exacerbated by the region's ongoing population pressures. Tapping new sources of water to meet the increased demand for fresh water would relieve some of the region's shortages, but as new sources of water become more expensive, they become less accessible to low-income countries, given those nations' limited financial and technical opportunities. At the same time, these low-income countries are often experiencing the fastest population growth in the region."

Add to the water shortages the world wide decline in arable farm land. In the next century, food and water are going to be more expensive. People can forgo expensive petroleum products, but food and water are the stuff of life. Historically, societies facing loss of these resources have responded with massive population migrations.

Greg Okin, professor of geography at the University of California at Los Angeles as quoted in Discover Magazine, May 2010:

"Desertification is happening today around the world, most notably in northern China, home to much of that nation's 1.2 billion citizens. The world needs more food, more land to grow it, and more water to irrigate it, yet we have the same amount of land, less water, and higher temperatures," ..."This is a train wreck about to happen that will impact hundreds of millions of people now and perhaps billions in the future, because that's how many live in dry lands worldwide."

I expect water wars by mid-century. Biological warfare might be a cost effective alternative to conventional invasion; providing a means to eliminate neighbors competing for scarce and valuable land and water resources without damaging those resources themselves. Many areas threatened by water shortage have a recent history of religious fanaticism. It seems reasonable to predict that the two will merge into mutually justified rationalization for warfare in multiple regions of the world, potentially causing world war three.

Can technology come to the rescue? What would Ray Kurzweil's world of artificial intelligence, robotics, genetic engineering, and nanotech do to alleviate this looming crisis? Water desalinization and greenhouse agriculture are both expensive. Nano technology and other scientific advances show much promise for capturing energy from the sun which can then be re-purposed to fresh water and food production. Genetic engineering has already produced commercial production of ethanol from modified bacteria. These advances are sure to accelerate. We will have many new options for food and water production. Bill Joy projects that this same nanotech or recombinant life is likely to accidently cause us unexpected problems and the cost of solving such problems must be factored in. In the undeveloped countries, many facing the most egregious shortages are also unfortunately those with some of the highest birthrates in the world. No one can predict how all these factors will interact in our future. I believe that complex systems tend toward the center; extremes being balanced and buffered by complex interdependencies; similar to today's debates about global warming.

In his just published book, *The World In 2050 — Four Forces Shaping Civilization's Northern Future*, Laurence C. Smith (vice chairman and professor of geography, and professor of earth and space sciences at UCLA) predicts *a* future 40 years from now *"in which global population has grown by nearly half, forming crowded urban clots around the hot lower latitudes of our planet."* He goes on to imagine, *"Mighty new poles of economic power and resources consumption have risen in China, India, and Brazil. People are urban, grayer, and richer. Many places are water-stressed, uninsurable, or battling the sea."* Laurence Smith expects that in 2050 more than half of the water consumed by drought beset humanity will be shipped to them from northern latitudes. Laurence Smith's goal in presenting his analysis was similar to mine in this book. He attempts to limit himself to outcomes *"deduced from big trends and tangible evidence already apparent today, rather than political ideology or* [his] *wonderful imagination."* In this book I have concentrated on the potential impacts of technology. Laurence Smith focused on the four global forces—demographics, resources demand, globalization, and climate change. Together we see a future where by 2050 humanity is dealing with numerous transformations of our world. We will be unprepared! As Ian Morris summarized *"we need to keep the dogs of war on a leash, manage global weirding, and see through a revolution in energy capture…"* Humanity's poor track record as a steward of this earth leads me to suspect that during the last half of this century will see wars, starvation, disease, refugee flows, and a human population crash. Concurrent to these catastrophes we will also be advancing our sciences, and our technology; fumbling through the transformations of the singularity.

8 ARTIFICIAL INTELLIGENCE

Futurologists tend to look at the impacts of artificial intelligence, genetic engineering, nanotechnology and robotics in isolation. It is a simple matter to single out a technology and then make a linear regression projecting its future. The most commonly referenced current example is 'Moore's law'. Gordon Moore was originally speaking about IC complexity – which was applied by Dave House to computer performance, predicting it doubling every 18 months. Popular acknowledgment of Moore and House is that 'Moore's law' has successfully predicted the exponential growth of digital electronics over the past 45 years. Ray Kurzweil's book *The Singularity Is Near: When Humans Transcend Biology* has a theory of technology evolution that predicts the same exponential growth for modern economic and biological systems. This religious belief in technology's beneficial growth has been termed *"Apocalyptic AI"* by Robert Geraci, assistant professor of religious studies at Manhattan College in his book of the same name. He suggests that Apocalyptic AI is a technological faith that derived from Judaism and Christianity. It certainly takes religious faith to make predictions of technology's impact on humanity so detailed that Ray Kurzweil expects near term technology advances to enable his immortality. Unfortunately Ray's fervor to outlast a normal lifespan has eclipsed his more impressive insights about the three overlapping revolutions of genetic engineering, nanotechnology and robotics. I agree with Ray Kurzweil's postulate that machine intelligence, nanotechnology and organic biology will merge together to create synergistic results. It is this merger of these three disparate scientific disciplines that is thrusting humanity into a new age.

Prof. Dr. Hugo de Garis, recently retired from his role of Director of the Artificial Brain Lab (ABL) at Xiamen University, China. Ten years ago Hugo headed a team in Belgium, Europe tasked with designing the world's first artificial brain. At the time, he was concerned that he was building the first massively intelligent machine 'artilect'. He predicted that artilects would become infinitely smarter than human beings. In 2005 he wrote a book titled: *The Artilect War: Cosmists vs. Terrans : A Bitter Controversy Concerning Whether Humanity Should Build Godlike Massively Intelligent Machines*. Prof. Dr. Hugo de Garis's willingness to continue his work in China building artilects is an example of the ethical dichotomy frequently seen in technologists. They commonly feel comfortable developing potentially dangerous science as long as they have made society aware of the impending ethical choices. In counterpoint, this is the same dilemma that motivated Bill Joy to change his career in 2003, leaving his fabulously successful computer company, Sun Microsystems.

A more moderate view on the danger of artilects can be found in Jeff Hawkins recent book *On Intelligence*. Jeff Hawkins has spent his life studying both the human brain and computer science. His perspective on artificial intelligence is grounded in both biology and computers. Jeff Hawkins' book describes a comprehensive theory of how our brains work and then uses his theory to design computer intelligence emulating human thought processes. He has termed his theory Hierarchical Temporal Memory (HTM).

Jeff Hawkins has founded a company, Numenta Inc., to develop computer technologies that replicate the structural and algorithmic properties of the neocortex.

HTM offers the promise of building machines that approach or exceed human level performance for many cognitive tasks. Just a few years into development; HTM is already enabling low cost, generic, learning based systems with particular success at emulating human vision as applied to image analysis and object tracking. In the next chapter I will propose one dramatic potential of an HTM like computer which combines all elements of genetic engineering, nanotechnology and robotics. In the chapter *The Future Of Intelligence*, Jeff Hawkins proposes many fantastic, previously inconceivable applications of HTM in the next few decades. He counters the fears of Prof. Dr. Hugo de Garis' artilects causing social disruption. Jeff Hawkins questions whether intelligent machines will exhibit any characteristics capable of the 'Nightfall' scenarios quoted from Ian Morris in my chapter *Predestination*. His opinion is that intelligent machines are going to be one of the least dangerous, most beneficial technologies we have ever developed. I agree with his observation that there will be no reason to include the animal derived old brain instincts like fear, paranoia, and desire. Intelligent machines have no need for an analog of human emotions. Instead they will be dedicated and optimized for tasks like vehicle navigation or weather prediction.

As we move into the latter half of this century, brainlike machines will surpass all higher order human thinking capabilities. These artilects will have speeds a million times faster and capacities trillions of times larger than individual human brains. If you consider that the Internet only reached a critical mass in the early 1990s; in the last 20 years it has become one gigantic memory store for humanity's knowledge. With the geometric growth it will continue to experience, its eventual capacity is beyond human imagination.

Intelligent machines designed to tap into the global knowledge stores that supersede our today's Internet will have capabilities not imagined. How humanity is likely to interact with these vast computer intelligences is the topic of the next two chapters.

9 MAN/MACHINE

Computer scientist J. Storrs Hall's book on the development of artificial intelligence and its ethical aftermath, *Beyond AI: Creating the Conscience of the Machine*, says the following about our likelihood of proceeding rapidly down the path of merging man with machines:

> *"We've been modifying one another since the invention of speech, enhancing our memories since the invention of writing, enhancing our eyes with lenses, our muscles with machines, and our voices with electrical signals on wires. It is so ingrained in human nature for us to enhance ourselves that 'not' doing it would be the weird, inhuman thing to do."*

By the middle of this century we will be further supplementing our bodies with genetic, drug and electronic augmentations. Many futurists including J. Storrs expect inexpensive robots to be common in this time period. I'm in general agreement with this projection. My major caveat is that these will not be robot butlers and general purpose assembly workers with human like behavior. I believe that these robots will be purpose built appliances seamlessly integrated into our lives. A natural evolution of all the technology we take for granted today like cell phones, personal computers and high speed transit. Think for a moment how many electronic devices in your life fail within four years, are obsolete in two, and require more power than what they replace. This characteristic of the creations of the industrial revolution has not changed. In many ways the useful life of a device has shortened while the energy and materials required to design, build, and maintain them has increased. Nano technology and its power for self-assembly and self-repair has the potential to change all this. I think it will, but not in our current century. Consider that the human brain works all day on a couple of bananas, yet our current super computers require more than 50,000 times the same energy input. Consider that technology is getting so complex it is not profitable to repair or upgrade. Consider the constant battle for control of our electronic devices against cyber-attack; the attacks modifying as fast as biological viruses. Our defenses are always tardy and frequently ineffectual. Consider that a humanoid robot butler will not only have to have a superior general purpose computer to emulate a human, but must also have a power source to operate untethered, and must have physical properties that simulate the muscles and joints of the human body. It must operate for extended periods without failure of any critical component. It will have thousands of moving components susceptible to wear, and environmental degradation from moisture and grit.

Compare this to a biological butler; the result of many millions of years evolution optimizing its systems for operation up to periods of one hundred years. Humans use comparatively little energy, we are self-learning, self-repairing, adaptive, and creative. There are already six billion of us on the planet. There is no way that humanoid robots will be cost effective in this century. The biological alternative is orders of magnitude more efficient use of resources. What will happen instead is the augmentation of humans. Consider that with synergistic application of genetic modification, Nano technology, customized drug delivery systems, and brain-to-AI linkage we will have humans targeted for particular societal functions. The economics of this alternative are so dramatic that humanoid robots will not be competitive except for unique environments like radioactive areas or outer space. The primary roadblock to progress in augmenting humans is ethical. Storrs Hall's ruminations on this topic are erudite and extensive. Our ethical systems evolved to get human life to our current epoch. The most significant hurdle facing the human race is how to blend current philosophies developed over tens of thousands of years with incompatible ethics appropriate to future humans. The religious confrontations bedeviling society will be greatly exacerbated by augmented humans. It will be like creating millions of sociopaths. There won't even be commonality within the groups of augmented humans. Being modified purposefully, they will characterize new emerging social mores. Simply put, they will think differently. We don't have any analog of how people connected from birth in a hive mind will think. The hive mind will share advanced computer artificial intelligence. A hive seems very likely to engender new religions and atypically human ethical patterns.

The Pew Internet & American Life Project statistics showed that in 2010 half of teens send 50 or more text messages a day, or 1,500 texts a month, and one in three send more than 100 texts a day, or more than 3,000 texts a month. Two billion of the earth's population or one out of three people alive use the Internet. It is this generation that is going to be making decisions about using electronic implants to connect themselves to the artificial intelligence systems of the future. As J. Storrs is quoted at the beginning of this chapter, it would be atypical if large numbers of people from various cultures didn't enhance themselves. Once enhanced these people will immediately have extended intelligence, be more connected, and more productive than non-enhanced humans. The rudimentary beginning of the hive mind will be non-threatening. Just extension of what people are already doing on a frequent basis. As the biological connections mature to handle higher bandwidth and multiple senses; as the artificial intelligence of external connected computers improves; as the relations between connected individual deepens into new communication patterns we will discover that society has diverged. The young fully embracing these sweeping changes and the older generations resistant and further alienated.

10 ANDROIDS/BIOBOTS

As research work progresses on new frameworks of intelligence like Jeff Hawkins' Hierarchical Temporal Memory (HTM) we will soon begin to see artificial intelligence with specific capabilities exceeding humans. The Deep Blue chess computer that used massively parallel brute force computing to evaluate 200 million positions per second when playing chess with Gary Kasparov; did not exhibit learning or creativity. New artificial intelligence designs will be self-learning, able to creatively recognize patterns on top of patterns. These artificial intelligences will learn from experience. Early applications will be autonomous vehicles, medical diagnosis, and weather prediction. None of these uses exemplify Ray Kurzweil's three overlapping revolutions of genetic engineering, nanotechnology and robotics. I'm going to present what I consider a likely scenario that will be possible within fifteen years. This scenario will also lay the groundwork enabling the first generations of future humans.

Human genome DNA sequencing cost has steadily fallen. Within ten years there will be millions of fully sequenced people. In another decade whole populations will be sequenced, particularly in progressive homogeneous locations like Iceland. Initial uses of this data have been genetic screening and secure identification. Consider that this is the code of life. Unraveling that code will yield fundamental understanding of all life's processes. At this stage in our biological sciences this understanding is beyond our collective grasp. I believe it is within the grasp of an artilect.

What follows is a likely outcome starting in about 15 years. The enabling technology will be brainlike memory systems based on Hierarchical Temporal Memory (HTM), computers custom designed to support the HTM model, the lowered cost of human genome sequencing, and emerging within-body nano-sensor technology.

Imagine a new breed of artilect (I call it Genome-Brain). The primary data input for this self-teaching Genome-Brain is our genetic code. Our genome provides a natural pattern for the hierarchic self-learning input. A second sense (sense, in the model of sight-hearing-touch) would be less structured data. It would consist of medical records, intelligence tests, aptitude tests, education records, occupation history...associated with individual genomes. A third sense; instead of being fixed DNA patterns or static personal data records would be real-time biology data transmitted from nano-sensors within the bloodstream of a few thousand volunteers. The Human Metabolome Project has already identified 8,500 metabolites in humans. These cellular by-products will trace every chemical generated by cell metabolism in real-time. The Metabolome provides the biomarkers to associate gene expression with human bio-chemistry.

According to HTM theory, human creativity results from the mixing and matching of patterns in our neocortex. The cortical hierarchy is multi-level. Creativity is a result of abstraction of relationships of higher-order objects; recognition of patterns on top of patterns. After the 'Genome-Brain' has spent a decade accumulating billions of relationships between our genetic code, our real-time biology, and our resultant behavior, it will be able to identify specific groups of genes that control measurable traits like human intelligence. It will also be able to relate real-time biology with disease and genetic susceptibility. More interesting for the human race will be its capacity to begin to design new genetic material intended to modify the genetics of individual humans. Craig Venter's company Synthetic Genomics has already synthetically created the genome of a bacterium from scratch. In twenty-five years it will be common place technology to modify human DNA. The tough part will be to know what to modify. Genome-Brain will provide custom designed life forms to organizations like Synthetic Genomics. Feedback from the initial genetic designs (monitored by permanent nano-biosensors) will enable the Genome-Brain to refine its knowledge and enable improved genetic designs. Concurrently technologies for human computer implants and custom drug enhancements will complement these new genetic designs. For example, rudimentary computer interface brain implants will be co-evolved with genetic changes that enable improved communication between neurons and electronic brain implants.

Within fifty years the first examples of Technology Augmented Bioengineered (TAB) humans will be living among us. I've coined this acronym TAB to represent an intermediate form of human life, predecessors to the androids or biobots created towards the end of the decade. Unfortunately we humans will not be prepared for the social and ethical consequences. Prof. Dr. Hugo de Garis' warnings about the Artilect War are grounded in humanity's inability to rapidly evolve its social mores. After all, we are genetically programmed over two million years to enable our existing social structures. It is quite likely that the instincts enabling the growth of society were intrinsic to our success as a species. The disparity between genetically modified augmented humans will create massive social strife. The advantages of redesigned life will be so persuasive that some groups will ignore the consequences. Be it longer life, increased intelligence, or merging into the hive mind; the new humans will be so superior that jealousy and fear are the guaranteed reaction of the unenhanced multitude. This has been a popular topic for science fiction. It is no longer fiction. This will happen during our lifespans.

11 ECONOMICS

Many observers of human history have recognized cycles of behavior. Most of them have attempted to quantify these wave patterns of behavior hoping to be able to use them to predict the future. A recent interpreter of these cycles is John Casti who goes so far as to advance the science of Socionomics. He theorizes that cycles in human affairs are influenced more by social mood than specific events. In his book *Mood Matters – From Rising Skirt Lengths to the Collapse of World Powers*, John Casti argues the following postulates:

- *The long-term global mood has shifted from positive to negative;*
- *This will result in a massive change in every aspect of geopolitical, financial, and social life;*
- *The current problems are psychological, not financial.*

Analysts like George Friedman have suggested that the pattern of traditional life in developed countries is collapsing. He relates the disruption of historical social patterns more to the factors of longer life, the decline in fertility rates, extended education of children, and the breakdown of the institution of marriage. Jared Diamond points to over population and depletion of resources and warns that lifestyle of current human societies has less than 50 years left. Futurists more oriented towards technology like Martin Ford see the economic impacts of truly advanced future technologies as being the probable cause for social disruption in the next fifty years.

It is likely that these eminent authors and analysts have grasped fundamental truths. That humanity is approaching a Kuhnian paradigm shift is the common thread within all future predictions for the coming century. The point of this book is that we are facing more than a wave cycle. In my view our singularity will alter all the social patterns of past humanity and eventually humanity itself. The immediate disruption over the next few decades is the most easily visualized. Technologies such as robotics and artificial intelligence will change jobs and industries throughout the industrialized world. We've seen these types of social impact before. Jared Diamond points to the invention of subsistence agriculture 9000 years ago as the beginning of the human population explosion that is likely to peak in this century. The industrial revolution, less than 200 years ago moved humanity from the farm to cities. According to Kurzweil, his logarithmic graphs of 15 paradigm shifts for key historic events show an exponential trend. There are so many trends that are likely to impinge on the world over the next 50 years that successfully using them to predict the specific consequences on the economies of the world is not possible.

For every negative impact, technological revolutions have countervailing positive impacts. The luxurious life styles of the developed nations since World War II can be directly attributed to the productivity resulting from industrialization and technology. Because of the power of technology over the last two centuries we have consumed more resources worldwide, than humanity consumed in the rest of recorded history. This leverage of technology does not appear to be lessening. If Ray Kurzweil's logarithmic growth projections are accurate we will be receiving even greater benefits over the next several decades. Resources do not increase logarithmically. If anything, some of them have been decreasing logarithmically over the last century. Fish in the oceans, arable land, fresh water, and natural vegetation all are in dramatic decline. Technology's ability to introduce previously unidentified substitute resources will be enhanced. Humanity's historical shifts in fuel sources from forest trees, to coal, to fossil fuels, to nuclear power could not have been predicted by previous generations. What we can predict today is that there will be multiple new power sources that will continue to enable the technological world we have been building for ourselves.

How will this all impact the economic well-being of the citizens of the world? It seems likely that, though there will be severe regional economic disruptions, the overall trend will be the continued improvement of the living standards of undeveloped countries while still maintaining the living standards predominant in the developed world. The combined influence of genetic engineering, nanotechnology and robotics technologies will fuel this continued economic growth. However, the social disruptions of these same technologies combined with resource disparities insure that there will also be the specters of war, famine and disease causing regional societal collapse and human suffering.

During the latter half of this century as TAB humanity begins to emerge, the definition of social well-being will change. The definition of what it means to be human will be evolving. In a later chapter I address what this evolution of humanity will mean to our existing concepts of human equal rights. The definitions of social welfare and economic growth will be undergoing transitions as the rules governing human economics diverge into whatever will replace them in the post-humanity world.

12 WARS

Predicting that the human race has wars in its future requires no insight. We humans have never had a community that has built a lasting peace. The beliefs of Buddhism have come closest to achieving a path leading to the cessation of suffering. The Buddhists are particularly notable as the only major world religion to avoid fostering international war. Unfortunately despite examples of peaceful coexistence, humanity has not demonstrated much success at preventing armed revolution. The Brooking Institution's new report, *Poverty in Numbers: The Changing State of Global Poverty from 2005 to 2015*, states that:

> "By 2015, we will not only have halved the global poverty rate, but will have halved it again to under 10 percent, or less than 600 million people, with India and China responsible for three-quarters of the reduction in the world's poor expected between 2005 and 2015. While these findings likely come as a surprise to many, they shouldn't," says the report. "Growth lies at the heart of poverty reduction. As developing country growth took off in the new millennium, epitomized in the rise of emerging markets, a massive drop in poverty was surely to be expected."

As Jared Diamond has documented so carefully, the history of social development demonstrates repeated slow evolution in social maturity leading frequently to a rapid decline as available resources become constrained. The Brooking Institution's observation that poverty reduction is due primarily to emerging markets makes it clear that instead of humanistic teaching like Buddhism it is actually resources and technology that are the primary forces reducing suffering in the world. I quote Jared Diamond in this book's prolog as saying that the lifestyle of current human societies has less than 50 years left.

Jared Diamond's perspective is originally attributed to Thomas Malthus' *Essay on Population*, 1798, *"The power of population is indefinitely greater than the power in the earth to produce subsistence for man"*. The question in the modern world is how long the balance can be maintained with technology increasing the availability of resources while at the same time we consume ever greater quantities. Bucking the Malthusian trend of ever increasing population, the former Soviet Union and its allies are in regional population decline. Most of the developed world now has sub-replacement fertility levels. Within 20 years the average age of the developed world will be approaching 50. This aging demographic is occurring at the same time as we've past the point of sustainability in our consumption of worldwide resources. The Millennium Project's *2010 State Of The Future* predicts that 2 billion people are likely to be living in slums by the years 2050. *"Without sufficient nutrition, shelter, water, and sanitation produced by more intelligent human-nature symbioses, increased migrations, conflicts, and disease seem inevitable."*

These circumstances assure that future wars will be fought over resources. George Friedman suggests that the battlefield will be in outer space. Control of the high ground yielding the ability to make precision attacks anywhere on the Earth's surface. I've outlined a counter to this threat in my chapter on biological weapons. It detailed how a small group will be able to wreak havoc with genetically modified biological warfare agents. There are so many sources for conflict and means for attack that predicting the timing of future wars in specific regions is not possible. Instead I'd like to highlight two fundamental causes for regional conflict. One I've mentioned before is access to sufficient clean water, and the other is an entirely new source of conflict; control of Technology Augmented Bioengineered (TAB) technologies and DNA. I name the first category of conflict Water Wars, and the second, Bio-Wars. To quote again from *2010 State Of The Union*:

> *"Unless major political and technological changes occur, global water demand could be 40% more than current supply by 2030. This would cause conflicts over tradeoffs among agricultural, urban, and ecological uses of water, along with mass migrations and wars."*

Worldwide we will have soon depleted the water we have been mining from aquifers deposited during the last ice age. These are being sucked dry at the same time as global warming is likely to create persistent drought over major regions of the earth. Water wars are unavoidable. People can go forego electric power and petroleum. They can't forego clean water. Eighty percent of diseases in the developing world are water-related. These water wars cannot be avoided by the use of technology. All water purification and desalinization techniques are energy intensive, both in the manufacture of the technologies used, as well as in their implementation. No doubt in future centuries, TAB humans will have solved this problem. After all, 71% of the earth's surface is covered by water. Until then, this historical source of conflict will persist.

Bio-Wars…what could those be? Consider the following **hypothetical** scenarios. The year is 2050. During the two decades between 2010 and 2030 genetic engineering of human DNA progressed rapidly. Social response to this progress has fragmented the world with violent sectarian conflict along both religious and political lines. In response to Iceland's financial collapse in 2017, Decode Genetics, the gene-hunting company in Reykjavik, was nationalized by the Icelandic legislature. DeCode's access to the Icelandic Health Sector Database (HSD) containing the medical records and genealogical and genetic data of all Icelanders gave the company a head start in identification of the primary genes responsible for human intelligence. In a controversial move to assure Iceland's future, a population wide referendum persuaded Icelandic woman to act as maternal hosts, implanted with lab grown fetuses engineered for genetic brain enhancement. Feedback from preschool testing of their genetically enhanced offspring identified successful engineered strains of DNA allowing rapid improvements in Icelandic intelligence. The average IQ of Icelandic teens is now measured at 210. More than a hundred percent increase in just 40 years. The entre of Icelanders with augmented intelligence into the Icelandic workforce has already given Iceland the highest living standard in the world. Control of Decode Genetics' genetic engineering database is considered an Icelandic national resource and is protected by many layers of secrecy and military security.

Elsewhere in the world: Kim Jong-un has augmented North Korea's nuclear deterrent developed by his father Kim Jong-il, with an army of genetic clones. As the last Stalinist state on earth, Kim Jong-un had required some means to combat desperation inside his country, and resentment toward the Kim dynasty. To accomplish this Kim repurposed the massive underground war facilities beneath Pyongyang. Instead of tunnel warfare these now provide accommodations for North Korea's new army. For the last 40 years genetic experiments hidden underground have been protected from scrutiny. Kim Jong-un's secret army was revealed this year because he used it to suppress a widespread revolt started by students at Kim Il-sung University. Fearing civil war Kim unleashed his clone army. Popular support for the students withered as science fiction clone warriors suppressed the slightest protest with deadly force. Apparently these engineered human clones are physically superior to the finest existing athletes. They feel no pain. They see in the dark. They act with no moral constraints. When injured, nanobots in their blood seal wounds in seconds, making clone warriors very difficult to incapacitate with bullet wounds. Universal horror, at these events, is being voiced both within North Korea and around the world. South Korea's conventional army has asked for the United Nations to censure North Korea. International debate rages over whether North Korea's cloned warriors are protected as combatants as defined by international humanitarian law. The US defense department has joined with the CIA to discover the bioengineering secrets behind North Korea's new armed forces.

Both these scenarios may appear unlikely to you. To change your perspective, consider how threated the community you live in now would feel if your neighboring city was composed of people twice as intelligent as yours? I would expect universal concern. The threat is not just towards you but to the potential success of your children and grandchildren. Instinctual response for species survival would infect your town with all the xenophobic emotions that typify the worst of man. This reality of superior neighbors is much more threatening than if the nearby city varied from yours by only ethnic or religious differences?

The Crusades of the 11th, 12th, and 13th-centuries were fundamentally fought over economics. What motivations sufficed to keep the wars going for 200 years? The European indignation about the 3000 Christian Pilgrims massacred in Jerusalem was enough to muster the original popular support. Crusaders of the upper classes saw an opportunity for acquiring fame, riches, lands, and power.

If the Christian Europeans had believed that the Turks' civilization and the religion of Islam were superior; and on the verge of dominating their own...how long would the conflict have persisted? Likely until one side or the other had committed complete genocide against the other. The conflict between our two hypothetical towns differing only in intelligence might also be compared to the Spanish conquest of the Inca Empire. Francisco Pizarro's advantages were steel weapons, gunpowder and smallpox. With a force less than 200 men, Pizarro subdued the Inca civilization. What advantages will a society accrue if all its members have beyond human mental capacities? Their chance of inciting attack from their neighbors would undoubtedly increase if they were practicing Buddhism. A non-violent super intelligent society would be a particularly tempting target for attack by neighboring communities embolden by their willingness to use force of numbers and lethal weapons.

Human experience has been an endless battle between the strong and the weak. Social systems and genetically derived human instincts will not have time to evolve to another pattern during the brief period of this century. Technology Augmented Bioengineered (TAB) humans are inevitable. The immediate consequence must certainly be human conflict over the very definition of future intelligent life on this planet. I called this threat **Bio-Wars**. It is threat of TAB life so different that it will rapidly supplant humanity as we know it. This threat is the threat of extinction. As such, it will engender wars of species survival.

13 INTERPLANETARY SPACE

The burden of human and future human 'TAB' life on our planet's natural resources assure that they will eventually be depleted below our collective needs to consume them. An unbiased observer from outer space might even consider humanity to be a plague species, permanently undermining the health of its host planet. The question for this chapter is to surmise how likely a hypothetical outer space intelligence is to meet a human? The high probability of the existence of other intelligent life within our universe is heavily predicated on the vast size and age of the universe. This same metric makes the likelihood of them meeting humankind vanishingly small. We have existed for less than a blink of an eye in cosmic time. Much more likely is that alien intelligent life will meet our machine offspring (or earth sourced biology) custom engineered for life on other celestial bodies.

Impartial analysis of our human biology is that we are evolved to live under very specific and limited environmental constraints. Change the atmosphere, the gravitational force, the exposure to radiation, the requirement for water and carbohydrate food sources, even the loneliness of isolation, and we fail to thrive. Bacteria survive extended exposure to deep space. We don't last even a minute. The logistics for maintaining delicate human life in near earth orbit has proven to be prohibitively expensive. Compared to this, NASA's Explorer Missions have been in operation for decades. The longest of which, (IMP 8) has been operational for over 26 years and still produces valuable information about the solar wind.

Space exploration by even the earliest versions of robotics has proven to be both cost effective and robust. As the order of magnitude improvements in robotics and artificial intelligence progress over the next decades it will become glaringly obvious that space exploration is the realm of the machine. Assuming that we don't discover a mechanism for faster than light space travel in the next few centuries, the only human biology that will be sent to habitable planets is our DNA code. It won't be our DNA specifically. It will be genetic instructions derived from human genes custom designed as enhanced versions of intelligent life suited for planetary habitation. This manufactured life will likely be a conceptual ancestor of the selfish machine (Richard Dawkins, *The Selfish Gene*) programmed to do whatever is best for its genes as a whole. TAB humanity will have already been superseded by the cyborg life forms envisaged in popular science fiction. Self-maintaining factory spaceships will carry and protect the chemicals necessary for building biological life the hundreds of light years required for traversing cosmic distances. Once parked in orbit the factory ship will send down robotic systems designed to create a habitat for birthing life on the target planet. Computer stored genetic code will be used to manufacture new life from scratch. Trials of the early pre-planned species will be tested in the new planet's ecosystem. Results of these trials will hone the suitability of the customized life for unexpected environmental constraints.

The eventual destiny of intelligent biologic life being sent to other planets will depend on the continued success of intelligent life on Earth. Predicting the results of the co-existence of TAB humanity, cyborgs, robots, and artificial intelligence is not possible. It is plausible to predict that biologic life will be sent out repeatedly to destinations outside our solar system. In the oncoming millennium, DNA carried by spacecraft from Earth is the most likely source for carbon based life in our solar system.

In his book, The Next 100 Years: A Forecast for the 21st Century, George Friedman predicts that technological progress in this century will focus on space for both military strategy and solar power. For reasons identified earlier in this chapter I believe that the dominate space faring community will be intelligent robots. They will be both autonomous and also guided by Earth based control. The competition for the strategic military high ground will be the impetus for rapid evolution of these systems. The likely utility of a space elevator to cost effectively ferry material up to geosynchronous orbit would seem to be the most promising vehicle for near Earth space development. Such elevators will be very time consuming to develop and launch. Unfortunately they would be tremendously susceptible to military or terrorist attack. If damaged the carbon nanotube based elevator cable would whip down circling the earth many times with destructive force. For this reason I think that space elevators will have to wait until the Earth is under some manner of central control.

I also have doubts of satellite based solar collectors beaming to Earth microwave or laser energy. The same critique applies. These power plants would be tremendously fragile; susceptible to micrometeorites, solar storms, and military attack. The potential cost effectiveness of space based solar energy capture versus the rapidly advancing engineering for land based energy capture has yet to be demonstrated. My favorite is nanotechnology or engineered photobacteria directly converting solar radiation to hydrogen gas. I doubt that space based power will be a significant component of Earth based consumption for many centuries.

14 EQUAL RIGHTS

Modern humanity separates itself from other life forms (and earlier human history), in part, by glorification of our capacity to codify and enforce equal rights for man. From a religious view point these modern ethics practiced in many developed countries must give homage to all major religions. In particular, the Buddhist goal for the cessation of human suffering is of a much more generous spirit than Christianity or Islam, which required belief in their respective gods. As I mentioned in my chapter on human instincts; morality is hard wired into our brain conferring evolutionary advantage. Storrs Hall's book, *Beyond AI: Creating the Conscience of the Machine*, observes that, *"After millennia of philosophical investigation, we have only just begun to realize that our morals, too, arise from our evolutionary origins."* Storrs Hall exhaustively explores what ethics are fundamental to what it means to be human and what kind of ethics we might expect from machine intelligence.

Aspects of social cognition have been documented in primates, and corvids. Nathan Emery and Nicola Clayton published a paper in *Science* (2004); The Mentality of Crows: Convergent Evolution of Intelligence in Corvids and Apes. They argue that cognitive abilities evolved multiple times in distantly related species with vastly different brain structures in order to solve similar socioecological problems. Storrs Hall theorizes that ethics evolve in intelligent species under selective pressure that rules against individual interest to the advantage of the group. Storrs Hall presents logical arguments why some of these evolutionary pressures will apply to intelligent machines. Vernor Vinge is famous for his presentation at the VISION-21 Symposium sponsored by NASA Lewis Research Center, March, 1993. He began his presentation thus: *"Within thirty years, we will have the technological means to create superhuman intelligence. Shortly after, the human era will be ended."* Vernor Vinge, Ray Kurzweil, and Storrs Hall all wonder if there is the possibility for co-existence between humanity and intelligent machines. It seems obvious to me that this question is only of importance in the current century. In the preceding chapters I have discussed the myriad pressures that will come to bear on humanity over the next 90 years. A beneficent perspective would have society guiding the development of TAB humans and artificial intelligence so that the rampant disparities emerging between humans, between generations, between societies, between regions, between (humans & TAB humans & robots & artificial intelligence); would treat those less privileged with humanity. Pretty unlikely!

Much more likely is a frightful, rapid and messy evolution into multiple new intelligent species. Sometime after the year 2100, the emerging biobots, robots, and massive artificial intelligences will take over control of their own evolution. At this point the discussion of equal rights will be out of humanity's hands. How superhuman intelligence decides to treat the remnants of humanity will have more in common with how they decide to treat crows (Corvids). Biologically derived social behavior is shared by crows, primates and humans. The morality of biobots and machine life will be dealing with different motivations to solve different survival imperatives. The equal rights of super-intelligent life forms have yet to be written and we won't be around to critique them.

15 RELIGION

As the brief last chapter topic in this book, you as the reader may wonder why I'm dealing with religion now. The answer is that all the previous chapters impinge on the topic of religion. Religion has been with humanity since before its earliest recorded history. Based on artifacts found in Neanderthal Man burial sites, it is now considered possible that they believed in an afterlife. Surely the capacity for faith is deeply imbued within our genome. In his recent treatise, *The Faith Instinct*, on the history of religion and why it has been essential to human success, Nicholas Wade points out that the role religion plays in shaping human evolution is far from understood. Nicholas Wade questions, "*Is there not some way of transforming religion into versions better suited for a modern age?*" He concluded his book with the following paragraph:

> *Maybe religion needs to undergo a second transformation, similar in scope to the transition from hunter gatherer religion to that of settled societies. In this new configuration, religion would retain all its old powers of binding people together for a common purpose, whether for morality or defense. It would touch all the senses and lift the mind. It would transcend self. And it would find a way to be equally true to emotions and to reason, to our need to*

belong to one another and to what has been learned of the human condition through rational inquiry.

In the chapter on Human Instincts I introduced the concept of a hive mind, a collective consciousness. The Internet, cell phone texting, and the implantable brain computer interface have brought us to the brink. The only step remaining is for TAB humans to share wireless communication. This capacity is currently being funded by DARPA for brain transducers worn in an external helmet. The constraint against using the much more effective implanted electronic communicator is more political than technical. Within a generation many young people will have become part of the emerging hive mind. The central hypothesis of Socionomics is that the flow of social behavior and action is defined by the formula: Herding Instinct → Social Mood (beliefs/feelings) → Social Behaviors and collective events. John Casti's theory is that societal mood impacts the character and timing of all human events. If this theorem is true then how does the instantaneous sharing of that social mood impact our future? The ramification of TAB humanity connected by a hive consciousness which includes all recorded knowledge, ultra-intelligent computers, and trillions of remote sensors is beyond boggling. What will these future humans use to bind themselves into a community, to protect their independent souls from the collective, to explain the still unanswered questions about the universe? I believe this offers a possible agenda for a future religion; a religion that accepts both the individual and the collective. A religion that instills the new forms of morality required for future intelligent life.

Simon Conway Morris has written extensively about convergence having guided natural selection. He argues convincingly that the direction and inevitability he believes is inherent in evolution follows universal physical laws. He explains how complexity on the large scale arises from simple laws on the small scale and why humanity may not be the result of a random evolutionary process. If this theory is correct then intelligent life is certain to be found throughout our universe. The transition from evolutionary biology to manufactured intelligent life is sure to be common place. What part religion may play in a rational world comprised of a multitude of intelligent life forms cannot be either discounted or defined. Social behavior is always the result of a context. There is certain to be complex social behavior in the emergent world of biobots and artificial intelligence. If this new pattern of behavior includes what we call religion, remains to be discovered.

16 SUMMARY

George Friedman has observed that the entire pattern of traditional life is collapsing and no clear alternative patterns are emerging. This view is appropriate to our times. The human race is in transition. No one; human or artificial intelligence can predict the outcome of our next century. What can be predicted is the evolution of a number of technologies whose inertia makes them unstoppable. Genetic engineering, nanotechnology and robotics are the harbingers of a very different community of intelligent life forms on earth. With the addition of artificial intelligence these four technologies will be the basis for the intelligent life that continues to evolve on Earth and within our solar system. Few people alive today will look on this future with anticipation. I doubt that even Ray Kurzweil will continue to believe in immortal man; even if he does achieve that milestone in this century. The Universe that we know is an environment without an example of immortality. Even stars are in a constant evolution. Stellar evolution starts with a molecular cloud and ends with a planetary nebula restarting the cycle of a stellar life. All levels of the known universe demonstrate a continuous fight for resources. The microcosm of life's evolution on Earth is no different. The Technology Augmented Bioengineered (TAB) humans emerging from this twenty first century will be competing between themselves; as well as competing with newly emergent forms of non-biological life. An immortal mind derived from Homo sapiens will be a museum piece. It will not have the knowledge, mental capacity or speed of thought to even usefully communicate with the intelligences emerging from this century. Proponents of virtual immortality imagine that they will have augmented brains able to maintain their humanity while ensconced in an artificial brain. This egocentric concept is an outcome from our old lower brain. All of our instincts and emotions evolved over millions of years to create the soul of modern man. These instincts and emotions are not knowledge. They are essentially the same in all humans. As such, they could be replaced by a single identical subsystem in an artificial brain attempting to replicate a human; but why? These instincts and emotions were important for survival in our past evolutionary biology. They will be counterproductive in artificial intelligences, robots, and eventually suppressed or discarded in TAB humans.

Homo sapiens have existed for only the last 500,000 years. In contrast, early protoctist life is 590,000,000 years old. Primates have lived on earth for more than 16 million years. Our anthropocentric optimism about the greatness of human civilization has little to substantiate it. Seen in the perspective of life's evolution on our planet, Homo sapiens will have one of the shortest life spans of intelligent species. There is little possibility that we will ever reach the mean lifetime of mammal species. Many of these will outlive our species by millions of years.

What should you, I and our children do to prepare for this new world? A tough question. It is easier to identify what not to do. Fighting the forces of change didn't work for the Luddites and won't work for modern fundamentalist religions. Having children, raising them, and socialization with neighbors and relatives are the greatest functions of the Amish family. This traditional pattern of family life is collapsing. The Amish, Islam and similar groups will find themselves so isolated from future societies that they will share no common values. This disparity will make it hard to persist the traditional social patterns of rural life. Within a few more generations it will be difficult to see any difference between the Amish and any other aboriginal culture. These sidelined communities will endure, hopefully without persecution. However, they will have given up their option to contribute to the future of humanity. Given that the majority of the world continues to have faith in a creator it is certain that some of these religions will develop beliefs that encompass TAB humanity as part of God's plan. For those that don't, the consequences of intolerance of change will inevitably lead to massive social conflict. As I said in the prolog, we are beginning a frightful, rapid and messy evolution into multiple new intelligent species. Conflict is inevitable. As the last few generations of humanity we will constantly be faced with choices for which our evolutionary evolved skills have not prepared us. No choice is left to maintain our old patterns of life. We are on a steep slope sliding towards extinction.

ABOUT THE AUTHOR

Alan Hoshor is a long time resident of the Pacific Northwest. Originally trained in Forestry at University of California, Berkeley; he has had a very diverse and eclectic career in industries both large and small. Twelve years ago he started Hoshor Systems Consulting to service small businesses in Washington State. Alan's claim to competence is his capacity for systems thinking matured over decades of productive life.

www.ingramcontent.com/pod-product-compliance
Lightning Source LLC
Chambersburg PA
CBHW060216290526
45789CB00003B/1282